Richard Evan Day

Electric Light Arithmetic

Richard Evan Day

Electric Light Arithmetic

ISBN/EAN: 9783337249151

Printed in Europe, USA, Canada, Australia, Japan

Cover: Foto ©berggeist007 / pixelio.de

More available books at **www.hansebooks.com**

ELECTRIC LIGHT ARITHMETIC.

ELECTRIC LIGHT
ARITHMETIC

BY

R. E. DAY, M.A.

EVENING LECTURER IN EXPERIMENTAL PHYSICS
AT KING'S COLLEGE, LONDON

London
MACMILLAN AND CO., Limited
NEW YORK: THE MACMILLAN COMPANY
1897

RICHARD CLAY AND SONS, LIMITED,
LONDON AND BUNGAY.

First printed 1882. *Reprinted* 1883, 1887, 1888, 1889, 1891, 1893, 1897.

PREFACE.

SINCE the year 1878, when I undertook the direction of the Evening Classes in Physics at this College, I have sedulously encouraged the working out by the students themselves of numerous arithmetical examples having a direct bearing upon the particular course of experimental lectures which, at the time, they are attending, and I find that by doing so they rapidly acquire a firm grasp of the principles of the subject, as well as a knowledge of its details, which it would be almost impossible for them to get by any other means.

That the students themselves are conscious of this is, I think, established by the fact that within less than four years the attendance at our Evening Classes in Physics has increased more than fourfold.

We have naturally had to pay a good deal of attention of late to the principles and practice of Electric Lighting, and it has been suggested to me that the arithmetical examples on this subject, which I have drawn up for the use of my own classes, might be of some service to a wider circle of students.

In compiling these problems I have always had to keep in view the fact that the majority of our evening students are engaged during the daytime at other pursuits, so that they cannot devote much time to going deeply into the subject, and also that as a rule they have but a slight acquaintance with mathematics, so that no examples must be introduced which require for their solution anything beyond decimal fractions and elementary algebra.

In the statements of the problems and in their solutions I have not thought it necessary to go into minute details about electrical formulæ and theories because these will be found in all the recognised text-books on the subject, and this collection of examples is not intended to replace but to supplement whatever text-book on electricity the student may be using.

R. E. DAY.

KING'S COLLEGE,
  June, 1882.

CONTENTS.

	PAGE
ELECTRICAL RESISTANCE OF WIRES AND LAMPS . .	1
INTENSITY OF CURRENT IN A SIMPLE CIRCUIT . .	21
HEATING EFFECT OF THE CURRENT IN A SIMPLE CIRCUIT	30
WORK UTILISED IN A SIMPLE CIRCUIT	45
COMPOUND ELECTRICAL CIRCUITS	57
DISTRIBUTION OF ENERGY IN A COMPOUND CIRCUIT	69
TABLE OF SQUARES, SQUARE ROOTS, AND RECIPROCALS	83
TABLE OF THE BIRMINGHAM WIRE GAUGE	88

ELECTRIC LIGHT ARITHMETIC.

ELECTRICAL RESISTANCE OF WIRES AND LAMPS.

THE electrical resistance (R) of a conductor varies directly with its length (L), and inversely as its sectional area (S), and may be expressed in terms of these quantities by the equation,

$$R = a\frac{L}{S},$$

where a is a constant quantity depending upon the material of the conductor and is numerically equal to the resistance of a wire of this material whose length and cross section are each unity. It is called the *specific resistance* of the material.

Hence if we have two wires whose lengths are l_1 and l_2, their sectional areas s_1 and s_2, their specific resistances a_1 and a_2, and their actual resistances R_1 and R_2, we shall have

$$R_1 = a_1\frac{l_1}{s_1} \text{ and } R_2 = a_2\frac{l_2}{s_2},$$

whence by division we get

$$\frac{R_2}{R_1} = \frac{a_2}{a_1} \times \frac{l_2}{l_1} \times \frac{s_1}{s_2};$$

and by means of this equation all problems connected with the relative resistances of wires in simple circuit can be readily solved.

Example 1. If the resistance of one mile of a certain electric light cable be 3·58 Ohms, what is the resistance of 3·7 miles of the same cable?

If we apply the data of this question to the above formula we find that

$$\left.\begin{array}{l} s_1 = s_2 \\ a_1 = a_2 \\ l_1 = 1 \\ l_2 = 3\cdot7 \\ R_1 = 3\cdot58 \end{array}\right\} \text{ and } \therefore \text{ the equation becomes } R_2 = 3\cdot58 \times 3\cdot7 = 13\cdot25 \text{ Ohms.}$$

Example 2. If the resistance of 700 yards of a certain cable be 0·91 Ohm, what will be the resistance of 1320 yards?   *Ans.* 1·72 Ohm.

Example 3. The resistance of 100 yards of a certain wire is equal to 5 Ohms, what length of the same kind of wire would have a resistance of 13·2 Ohms?

In this case the quantity whose value is required is l_2, and therefore transposing the equation (on page we have

$$l_2 = l_1 \times \frac{R_2}{R_1} \times \frac{a_1}{a_2} \times \frac{s_2}{s_1}.$$

But in the present case $a_1 = a_2$ and $s_1 = s_2$ and the equation reduces to

$$l_2 = l_1 \times \frac{R_2}{R_1},$$

and from the data of the question $l_1 = 100$, $R_1 = 5$, $R_2 = 13\cdot 2$.

Therefore

$$l_2 = 100 \times \frac{13\cdot 2}{5} = 264 \text{ yards.}$$

Example 4. The resistance of a certain cable is found to be 4·55 Ohms and the resistance of a mile of this cable is known to be 1·3 Ohm. What is its length?
Ans. 3·5 miles.

Example 5. If the resistance of 130 yards of copper wire 1-16th of an inch in diameter be one Ohm, what is the resistance of the same length of copper wire 1-32nd of an inch in diameter?

Since the areas of circles are proportional to the squares of their diameters we have

$$\frac{s_1}{s_2} = \left(\frac{32}{16}\right)^2 = 4;$$

also

$$\left. \begin{array}{c} a_2 = a_1 \\ l_2 = l_1 \\ R_1 = 1 \end{array} \right\}$$

and therefore the fundamental formula reduces itself to

$$R_2 = R_1 \times \frac{s_1}{s_2} = 4 \text{ Ohms}.$$

Example 6. What is the resistance of a mile of copper wire which has a diameter of 65 mils if the resistance of a mile of copper wire 80 mils in diameter is 8·29 Ohms? *Ans.* 12·56 Ohms.

N.B.—A *mil* is the one-thousandth part of an inch.

Example 7. What is the diameter of a copper wire a mile long which has a resistance of 23 Ohms if a mile of copper wire 70 mils in diameter has a resistance of 10·82 Ohms?

Let d_1 and d_2 represent the diameters of the wires in mils, then we have

$$\frac{s_2}{s_1} = \left(\frac{d_2}{d_1}\right)^2$$

and

$$\therefore \frac{d_2}{d_1} = \sqrt{\frac{s_2}{s_1}};$$

and from the data of the question

$$\frac{s_2}{s_1} = \frac{R_1}{R_2};$$

whence

$$d_2 = d_1 \sqrt{\frac{R_1}{R_2}} = 70 \sqrt{\frac{10 \cdot 82}{23}} = 48 \text{ mils.}$$

Example 8. A length of a thousand feet of wire 95 mils in diameter has a resistance of 1·15 Ohm: what is the diameter of a wire of the same material of which the resistance of 1,000 feet is 10·09 Ohms?

Ans 32 mils.

Example 9. Find the resistance in Ohms of 500 yards of copper wire 165 mils in diameter, the resistance of one mile of pure copper wire 230 mils in diameter being equal to one Ohm.

Since

$$\frac{s_2}{s_1} = \left(\frac{d_2}{d_1}\right)^2$$

the fundamental formula becomes

$$R_2 = R_1 \times \frac{l_2}{l_1} \times \left(\frac{d_1}{d_2}\right)^2 = \frac{500}{1760} \times \left(\frac{230}{165}\right)^2$$

$$= 0{\cdot}55 \text{ Ohm.}$$

Example 10. If we take as our unit of electrical resistance that of a metre of pure copper wire one millimetre in diameter, what will be the number expressing the resistance of a copper wire 6·358 feet long and 0·0336 inch in diameter, assuming that one foot = 0·3048 of a metre? *Ans.* 2·66.

Example 11. A copper wire 6 metres long is found to have a diameter of 0·74 of a millimetre; what will be the length of a copper wire of one millimetre diameter which will offer the same electrical resistance?

Since $R_2 = R_1$ and $a_2 = a_1$, the fundamental equation becomes

$$l_2 = l_1 \times \left(\frac{d_2}{d_1}\right)^2 = 6 \times \left(\frac{1}{\cdot 74}\right)^2 = 10{\cdot}957 \text{ metres.}$$

Example 12. What length of copper wire 4 millimetres in diameter would be equivalent to 12 yards of copper wire one millimetre in diameter?

Ans. 192 yards.

ELECTRIC LIGHT ARITHMETIC.

Example 13. Find the resistance of 15 miles of iron wire 0·3 inch in diameter having given that the resistance of one foot of iron wire 0·001 inch in diameter is 59·1 Ohms.

Since $a_2 = a_1$ the fundamental formula becomes

$$R_2 = R_1 \times \frac{l_2}{l_1} \times \left(\frac{d_1}{d_2}\right)^2 = 59 \cdot 1 \times \frac{15 \times 1760 \times 3}{(300)^2}$$

$$= 52 \text{ Ohms.}$$

Example 14. The resistance of 47 feet of copper wire 22 mils in diameter being 1 Ohm, find the resistance of 200 yards of copper wire 134 mils in diameter.   *Ans.* 0·34 Ohm.

Example 15. What must be the length of an iron wire the area of its cross section being 4 square millimetres, if it is to have the same electrical resistance as a wire of pure copper 1000 yards long whose sectional area is 1 square millimetre, taking the conductivity of iron equal to 1–7th of that of copper?

Since in this case the value of l_2 is required, we transpose the fundamental formula thus

$$l_2 = l_1 \times \frac{R_2}{R_1} \times \frac{s_2}{s_1} \times \frac{a_1}{a_2},$$

8   ELECTRIC LIGHT ARITHMETIC.

and from the data of the question $R_2 = R_1$; $a_2 = 7a_1$; $s_2 = 4s_1$ and $l_1 = 1000$. Substituting these values we get

$$l_2 = \frac{1000 \times 4}{7} = 571\frac{3}{7} \text{ yards.}$$

Example 16. How thick must an iron wire be so that for the same length it shall offer the same electrical resistance as a copper wire 2·5 millimetres in diameter?
Ans. 6 61 millimetres.

Example 17. If the specific resistance per cubic centimetre of a certain metal be 13·36 Microhms, what will be the resistance of a wire of this material one metre long and two millimetres in diameter?

N.B.—The Microhm is equal to the millionth part of an Ohm.

Referring to the definition of specific resistance given on page 1, we have in the present case

$$a_2 = a_1; \; l_2 = 100; \; l_1 = 1; \; s_1 = 1; \; s_2 = \frac{\pi}{100};$$
$$R_1 = 13\cdot36,$$

and substituting these numerical values in the fundamental equation

$$R_2 = \frac{13\cdot36 \times 100 \times 100}{\pi} = 42526 \text{ Microhms.}$$

Example 18. A wire one foot long and one mil in diameter has a resistance of 9·15 Ohms: find the specific resistance per cubic inch of the material of this wire.   *Ans.* 0·5989 Microhm.

Example 19. Find the resistance of 20 yards of platinum wire 0·016 inch in diameter, the relative resistance of platinum with respect to copper being 11·3.

Referring to Example 14 we find that the resistance of 200 yards of copper wire 134 mils in diameter is 0·34 Ohm, and therefore for the quantities in the fundamental formula we have

$$R_2 = \cdot 34 \times \frac{20}{200} \times \left(\frac{134}{16}\right)^2 \times \frac{11\cdot 3}{1} = 26\cdot 95 \text{ Ohms.}$$

Example 20. What would be the resistance of 7 miles of iron wire 238 mils in diameter, the relative resistance of this particular iron with respect to copper being 7·5?   *Ans.* 49·79 Ohms.

Example 21. What should be the respective lengths of two wires of silver and lead so that they may each offer the same resistance as 10 inches of copper wire of the same thickness, the conductivity of silver and lead with respect to copper being 1·0467 and 0·0923 respectively?   *Ans.* 10·467 and 0·923 inch.

Example 22. What must be the relative thicknesses of wires of iron, silver, and platinum, of the same length, so that their resistances may be equal?

Ans. As 100 : 36 : 122.

Example 23. If the resistance of a wire 3 metres long and weighing 3 grammes be 5·88 Ohms, what is the specific resistance per cubic centimetre of the material, its specific gravity being 20·337?

If s_1 represent the sectional area of the wire expressed in square centimetres, then we have

$$s_1 \times 300 \times 20\cdot337 = 3$$

$$\therefore s_1 = \frac{3}{300 \times 20\cdot337}$$

of a square centimetre; also $l_1 = 300$; $s_2 = 1$ and $a_2 = a_1$; and therefore the fundamental formula becomes

$$R_2 = 5\cdot88 \times \frac{1}{300} \times \frac{3}{300 \times 20\cdot337} = \frac{5\cdot88}{300 \times 2033\cdot7} \text{ Ohm}$$

$$= 9\cdot638 \text{ Microhms.}$$

Example 24. Determine the specific resistance per cubic millimetre of a wire 437 millimetres long, which has a resistance of 0·1257 Ohm, and which weighs 0·411 gramme in air, and 0·365 gramme in water.

The weight of water displaced = ·046 gramme, and therefore the specific gravity of the material

$$= \frac{411}{46} = 8\cdot935,$$

and then following the method of the last example we get $R =$ 30·278 Microhms.

Example 25. Find the specific resistance per *cubic centimetre* of the material of which the above wire is composed.   *Ans.* 3·0278 Microhms.

Example 26. A wire 874 millimetres long and weighing 0·822 gramme in vacuo and 0·73 gramme in water, has a resistance of 0·1257 Ohm; find the specific resistance per cubic millimetre of the material.
Ans. 15·14 Microhms.

Example 27. A piece of silver wire 3 feet long was found to weigh 7·2 grains, and its resistance was 0·3026 Ohm. What would be the resistance of a wire one foot long weighing one grain?

If w_1 and w_2 be the weights of two wires of the same material whose lengths are l_1 and l_2 and their sectional areas s_1 and s_2, then

$$\frac{w_2}{w_1} = \frac{l_2 s_2}{l_1 s_1} \text{ and } \therefore \frac{s_1}{s_2} = \frac{l_2}{l_1} \times \frac{w_1}{w_2}$$

Substituting this value for $\frac{s_1}{s_2}$ in the fundamental formula, and remembering that $a_1 = a_2$, we have

$$R_2 = R_1 \times \left(\frac{l_2}{l_1}\right)^2 \times \frac{w_1}{w_2}$$

$$= \cdot 3026 \times \left(\frac{1}{3}\right)^2 \times \frac{7 \cdot 2}{1} = 0 \cdot 24 \text{ Ohm}.$$

Example 28. What are the relative resistances of two wires, one of which is 30·48 centimetres long and weighs 35 grammes, while the other is 18·29 centimetres long and weighs 10·5 grammes?

Ans. as 12 : 10, nearly.

Example 29. Find the resistance at 0° C. of 20 metres of German silver wire weighing 52·5 grammes, having given that the resistance at 0° C. of a wire of this material 1 metre long and weighing 1 gramme is 1·85 Ohm. *Ans.* 14·1 Ohms.

Example 30. If the resistance of a foot of pure copper wire weighing one grain be 0·2106 Ohm, and the resistance of a piece of ordinary copper wire 3 feet long and weighing 3·45 grains was found to be 0·5782 Ohm; compare the conducting power of this sample of wire with that of a similar one of pure copper.

If the second wire were of pure copper, then by the method adopted in the solution of Example 27 we should find that its resistance

$$R = 0\text{·}2106 \times \left(\frac{3}{1}\right)^2 \times \frac{1}{3\text{·}45} = 0\text{·}5494 \text{ Ohm.}$$

Therefore

$$\frac{\text{Conducting power of this wire}}{\text{\qquad ,, \qquad ,, \quad ,, a similar one of pure copper}} = \frac{5494}{5782} = \frac{95}{100} \text{ nearly.}$$

Hence it appears that the *conductivity* of this specimen of copper wire is about 95 per cent. of that of pure copper.

Example 31. It is found that 37 inches of a certain copper wire weighing 518 grains give a resistance of 0·0041 Ohm; find the conductivity of this copper compared with that of pure copper.

Ans. 94 per cent.

Example 32. Two samples of copper wire were brought to be tested, so a length of 20 feet was cut off each of them. The first weighed 150 grains and had a resistance of 0·613 Ohm, while the second weighed 164 grains and had a resistance of 0·547

Ohm. Find the percentage conductivity of each of these samples with respect to pure copper.

Ans. 91·6 and 93·9 per cent.

Example 33. The resistance of one mile of copper wire whose diameter was 0·065 inch was found to be 15·73 Ohms. The resistance of a wire of pure copper 1 foot long and 0·001 inch in diameter being 9·94 Ohms; find the percentage conductivity of the copper of the first wire.

Had the wire been of pure copper its resistance would have been

$$R = 9\cdot94 \times 5280 \times \left(\frac{1}{65}\right)^2 = 12\cdot42 \text{ Ohms,}$$

and

$$\therefore \frac{\text{conductivity of the given wire}}{\text{,, ,, a similar one of pure copper}}$$
$$= \frac{12\cdot42}{15\cdot73} = 0\cdot7896$$

and therefore the conductivity of the sample is about 79 per cent. of that of pure hard-drawn copper.

Example 34. An iron wire weighing 249 pounds per mile and having a diameter of ·135 inch is found to have a resistance of 24·14 Ohms per mile. Compare the conducting power of this wire with that of a similar one of pure copper.   *Ans.* 11·8 per cent.

Example 35. A certain kind of Brush lamp has a resistance of 6 Ohms, and there are 4 lamps in series which are separated from each other by spaces of 50 yards, and the dynamo-electric machine is 400 yards away from the nearest lamp. What must be the thickness of the leading wire if its conductivity is 96 per cent. of that of pure copper and the total resistance of the *lead* is 8 per cent of that of the lamps, no ground circuit being allowed?

Total length of leading wire $= 800 + 300 = 1100$ yards.

" resistance of lamps $= 4 \times 6 = 24$ Ohms.

" " " leading wire $= 24 \times \cdot 08 = 1\cdot 92$ Ohms;

and by Example 33 the resistance of a wire of pure copper 1 foot long and 1 mil in diameter $= 9\cdot 94$ Ohms.

But since d_2 is the quantity whose value is required, the fundamental formula becomes

$$d_2^2 = d_1^2 \times \frac{R_1}{R_2} \times \frac{l_2}{l_1}$$

where

$$R_2 = 1\cdot 92; \quad l_2 = 3300 \quad d_1 = \cdot 001;$$
$$R_1 = \frac{9\cdot 94}{0\cdot 96} \quad l_1 = 1$$

and therefore

$$d_2^2 = (\cdot 001)^2 \times \frac{9\cdot 94}{1\cdot 92 \times \cdot 96} \times \frac{3300}{1};$$

whence

$$d_2 = \cdot 001 \times \sqrt{\frac{994 \times 33}{1 \cdot 92 \times \cdot 96}}$$
$$= \cdot 133 \text{ inch,}$$
or about No. 10 B.W.G. wire.

Example 36. Ten Swan lamps, each of which had a resistance of 25 Ohms, were arranged in series, and the length of the leading wire was 50 yards. What was the diameter of the "lead," its conductivity being 80 per cent. of that of pure copper, if its resistance was 2 per cent. of that of the lamps?

Ans. 19 mils, or about No. 26 B.W.G.

Example 37. If the resistance of a certain incandescent lamp, which is 12 feet from the supply main, is 80 Ohms, and the leading wires are of copper whose conductivity is 85 per cent. of that of pure copper, what should be the diameter of the wire so that its resistance may be ·08 per cent. of that of the lamp?

Ans 66 mils, or about No. 16 B.W.G.

Example 38. Five Brush lamps, each having a resistance of 6 Ohms, are to be worked in series, The lamps are 40 yards apart, and the nearest is 200 yards from the dynamo machine. If 10 per cent. be allowed for loss of energy in the "lead," which is of pure copper, what must be its diameter?

ELECTRIC LIGHT ARITHMETIC. 17

By the question

$$\frac{\text{Resistance of (leading wire + lamps)}}{\text{,, ,, leading wire}} = \frac{100}{10}.$$

Therefore the resistance of leading wire = 1-9th of that of the lamps.

But resistance of lamps = 30 Ohms.

∴ resistance of *lead* = $\frac{30}{9}$ Ohms.

and length of the *lead* = 400 + 320 = 720 yds. Hence if d_2 = the diameter of the leading wire in inches we have as in Example 35,

$$d_2 = \cdot 001 \times \sqrt{\frac{9 \cdot 94 \times 9}{30} \times 720 \times 3} = \cdot 08 \text{ inch},$$

or about No. 14 B.W.G.

Example 39. Thirty-eight Brush lamps, each of which has a resistance of 6 Ohms, are to be connected together and to the driving machine by an electric cable 3·5 miles long. If one-tenth of the total energy in the external circuit is allowed for loss in transit and the conductivity of the "lead" is 90 per cent. of that of pure copper, what should be its diameter?

Ans. 90 mils.

c

Example 40. The legal Ohm is a resistance equal to that of a column of mercury one square millimetre in section and 106 centimetres in length at the temperature of melting ice. What is the resistance between its ends of a cylindrical column of mercury 2 metres long and 2 millimetres in diameter?

$$R = \frac{1}{\pi} \times \frac{200}{106} = 0\cdot6 \text{ Ohm nearly.}$$

Example 41. Siemens' old unit of resistance was that of a column of mercury one metre long and one square millimetre in section. What was this in legal Ohms? *Ans.* 0·9434 Ohm.

Example 42. The British Association unit of resistance being equal to 0·9889 of a legal Ohm, express a Siemens' unit in terms of a B.A. unit.
Ans. 0·9539 B.A.U. nearly.

Example 43. The resistance of the copper wire of a cable when the temperature was 12° C. was found to be 250 Ohms. What would be the resistance of this wire at 24° C., the variation in resistance of this sample of copper per degree Centigrade between these limits being approximately 0·39 per cent.?

If R_t and $R_{t'}$ be the resistances of a certain wire at the temperatures t and t' respectively, they may very approximately be connected by the relation

$$\frac{R_{t'}}{R_t} = \frac{1 + Kt'}{1 + Kt},$$

where K is a *constant* which has a particular value for each metal. In the present case we have

$$\frac{R_{24}}{250} = \frac{1 + 24 \times \cdot 0039}{1 + 12 \times \cdot 0039},$$

$$\therefore R_{24} = \frac{250 \times 1 \cdot 0936}{1 \cdot 0468} = 261 \cdot 2 \text{ Ohms nearly.}$$

Example 44. The resistance of a certain length of copper wire at 0° C. being 100 Ohms, what would it be at 15° C.? *Ans.* 105·85 Ohms.

Example 45. Some resistance coils made of platinum-silver alloy are correct at 12° C. What amount of error will arise from using them when the temperature is 30° C., the variation in resistance of this alloy for 1° C. being about 0·03 per cent.?

Here

$$\frac{R_{30}}{R_{12}} = \frac{1 + 30 \times \cdot 0003}{1 + 12 \times \cdot 0003} = \frac{1 \cdot 009}{1 \cdot 0036} = 1 \cdot 00538.$$

Hence the error due to change of temperature amounts to an increase of resistance of 0·538 per cent.

Example 46. The mean summer and winter temperatures in Shetland being 12° C. and 4° C. respectively, what would be the limits of error due to variation of temperature which might occur in using a set of platinum-silver resistance coils, accurately adjusted for the mean annual temperature?

Ans. ± 0·12 per cent.

Example 47. Between what limits of temperature could this set of coils be used without introducing temperature corrections, if the margin of error is not to exceed 0·25 per cent.?

Let x be the number of degrees above or below 8° C., for which the temperature correction would not be greater than 0·25 per cent. Then we should have

$$x \times \cdot 03 = \cdot 25. \quad \therefore x = 8 \cdot 3 \text{ nearly.}$$

Hence the required limits of temperature are approximately 16° C. and 0° C.

Example 48. What would have been the approximate limits of temperature if these coils had been made of *platinoid* wire, for which the temperature coefficient is 0·021 per cent.?

Ans. + 20° C. and − 4° C. approximately.

INTENSITY OF CURRENT IN A SIMPLE CIRCUIT.

All problems connected with this part of the subject are worked out by the aid of a formula which is known as Ohm's Law; namely

$$I = \frac{E}{R}$$

in which I represents the intensity of the current; E the resultant of all the electro-motive forces in circuit; and R the total resistance of the circuit.

Example 1. The internal resistance of a certain Brush dynamo machine is 10·9 Ohms, and the external resistance is 73 Ohms; the electro-motive force of the machine being 839 Volts. Find the strength of the current flowing in the circuit.[1]

In this case we have

$$E = 839 \; ; \; R = 73 + 10\cdot9 = 83\cdot9 \text{ Ohms,}$$

and therefore substituting these values in the above equation we get

$$I = \frac{839}{83\cdot9} = 10 \text{ Ampères.}$$

---

[1] In a rotating armature the resistance appears to vary with the speed, on account of the possession by every electric current of a property sometimes called *electric inertia*, and sometimes *self-induction*. In the following examples the *resistance* of the dynamo is to be understood to depend upon the particular conditions of the experiment.

Example 2. The resistance of the dynamo machine being 1·6 Ohm, and the external resistance 25·4 Ohms; calculate the strength of current in the circuit, the electro-motive force being equal to 206 Volts.

Ans. 7·6 Ampères.

Example 3. Three arc lamps in series have a resistance of 9·36 Ohms, while the resistance of the leading wires is 1·1 Ohm and that of the dynamo machine is 2·8 Ohms. Find what must be the electro-motive force of the machine when the strength of the current produced is 14·8 Ampères.

In this case we have

$$R = 2·8 + 9·36 + 1·1 = 13·26 \text{ Ohms,}$$
$$I = 14·8 \text{ Ampères;}$$

and therefore substituting these values in the equation

$$E = I \times R$$

it becomes

$$E = 13·26 \times 14·8 = 196·3 \text{ Volts.}$$

Example 4. A certain arc lamp has an electrical resistance of 2·5 Ohms, while that of the leading wires is ·5 Ohm, and that of the machine 0·5 Ohm. What must be the electro-motive force of the machine so as to send a current of 25 Ampères through three of these lamps and the leading wires?   *Ans.* 213 Volts nearly.

ELECTRIC LIGHT ARITHMETIC. 23

Example 5. Calculate from the following data the average resistance of each of three arc lamps which are arranged in series. The electro-motive force of the machine is 244 Volts and its resistance is 3·7 Ohms, while that of the leading wires is 2 Ohms, and the strength of current flowing through each lamp is 21 Ampères.

If x represent the average resistance in Ohms of each lamp, then the total resistance of the circuit is

$$R = 3x + 2 + 3\cdot7.$$

But by Ohm's law

$$R = \frac{E}{I}$$

$$\therefore 3x + 5\cdot7 = \frac{244}{21} = 11\cdot61 \text{ Ohms.}$$

$$\therefore x = 1\cdot97 \text{ Ohms nearly.}$$

Example 6. In an experiment with a Gramme machine working on a simple circuit it was found that the electro-motive force developed by the machine was 81·58 Volts, the strength of current 29·67 Ampères, and the external resistance 1·14 Ohm. What was the internal resistance of this Gramme machine? *Ans.* 1·61 Ohm.

Example 7. In a similar experiment with another dynamo machine its resistance was known to be equal to 4·58 Ohms, while the electro-motive force was 158·5 Volts, and the current 17·5 Ampères. What was the resistance of the exterior part of the circuit?

Ans. 4·48 Ohms.

Example 8. Three Maxim incandescent lamps were placed in series. The average resistance, when hot, of each lamp was 39·3 Ohms, and that of the dynamo machine and leading wires 11·2 Ohms. What electromotive force was required to maintain a current of 1·2 Ampère through the circuit?

In this case we have

$$R = 3 \times 39·3 + 11·2 = 129·1 \text{ Ohms,}$$

and $I = 1·2$ Ampère;

and therefore by Ohm's law

$$E = I \times R = 1·2 \times 129·1 = 154·9 \text{ Volts.}$$

Example 9. Two incandescent lamps, each having a hot resistance of 73 Ohms, were joined up in series to the poles of a dynamo machine, and the total resistance of the circuit was 158 Ohms. What was the electro-motive force of the machine when a current of 1 Ampère was circulating through it?

Ans. 158 Volts.

Example 10. The resistance of the arc of a certain Brush lamp was 3·8 Ohms when a current of 10 Ampères was flowing through it. What was the electro-motive force between the two terminals?

By Ohm's law

$$E = I \times R = 10 \times 3\cdot 8 = 38 \text{ Volts.}$$

Example 11. The electro-motive force between the terminals of a Serrin lamp was 31·1 Volts, and there was a current of 35·8 Ampères flowing through it. What was the resistance of the arc?

Ans. 0·869 Ohm.

Example 12. Twenty-five exactly similar galvanic cells, each of which had an average internal resistance of 15 Ohms, were joined up in series to one incandescent lamp of 70 Ohms resistance, and produced a current of 0·112 Ampère. What would be the strength of current produced by a series of 30 such cells through 2 lamps, each of 30 Ohms resistance?

The data of the first part of the problem enable us to determine the average electro-motive force of each cell of the battery. Let this be represented by E, then we have

$$25\ E = I \times R = \cdot 112 \times (25 \times 15 + 70)$$
$$= \cdot 112 \times 445$$
$$\therefore E = \frac{\cdot 112 \times 445}{25} = 2 \text{ Volts nearly.}$$

Then from the data in the second part of the problem we have by Ohm's law

$$I = \frac{30 \times 2}{30 \times 15 + 2 \times 30} = \frac{60}{510} = 0{\cdot}118 \text{ Ampère.}$$

Example 13. Two incandescent lamps whose resistances were 16·9 and 32 Ohms respectively, were joined up in series with a battery of 40 similar voltaic cells, the total resistance of which was 20 Ohms, and the strength of the current produced was 1·16 Ampère. What will be the strength of current produced by a battery of 60 such cells through a series of 4 lamps whose respective resistances are 16·9, 32, 20, and 16 Ohms?   *Ans.* 1·043 Ampère.

Example 14. What would have been the strength of current in Example 12 if the area of each of the battery plates had been doubled, all else remaining the same?

The resistance of a battery cell varies inversely with the area of the plates, and therefore we should have had

$$I = \frac{30 \times 2}{30 \times \frac{15}{2} + 2 \times 30} = \frac{60}{285} = 0{\cdot}215 \text{ Ampère.}$$

ELECTRIC LIGHT ARITHMETIC. 27

Example 15. What would have been the strength of current in the case of the four lamps in Example 13, if the size of the battery plates had been doubled?

Ans. 1·2 Ampère.

Example 16. At the Mansion House three Crompton arc lamps were joined up in series with a Bürgin dynamo machine. The resistance of each lamp was 2 Ohms, which was also that of the leading wires. The electro-motive force of the dynamo was 244 Volts, and the current 21 Ampères. What was the internal resistance of the dynamo machine?

If $x =$ resistance of machine in Ohms, then the total interpolar resistance being 8 Ohms the resistance of the circuit was $R = x + 8$, and substituting these values in the equation

$$R = \frac{E}{I},$$

we get

$$x + 8 = \frac{244}{21} = 11\cdot62$$

$$\therefore x = 3\cdot62 \text{ Ohms.}$$

Example 17. The current from a certain Brush machine, whose electro-motive force was 839·02 Volts, was sent through a series of 16 arc lamps each of which had a resistance of 4·51 Ohms. The resistance of the leading wires was 0·8 Ohm, and the strength of the current was 10·04 Ampères. What was the resistance of this Brush machine?

Ans. 10·61 Ohms.

Example 18. What was the difference of potential between the battery terminals in the case of the first battery described in Example 13?

As in the solution of Example 10 we have

$$P.D. = I \times R = 48·9 \times 1·16 = 56·724,$$
$$\therefore P.D. = 56·7 \text{ Volts nearly.}$$

Example 19. What would be the *P.D.* between the terminals in the case of the second battery mentioned in Example 13?

Ans. 88·55 Volts nearly.

HEATING EFFECT OF THE CURRENT IN A SIMPLE CIRCUIT.

Whenever an electrical current flows through a conductor, heat is invariably produced, and it can be proved theoretically, as also Joule has shown experimentally, that the total quantity of work converted into heat per second in a circuit may be expressed by the formulæ

$$W = I^2 \times R = \frac{E^2}{R} = E \times I \text{ ergs}$$

if the units be absolute

$$= E \times I \times 10^7 \text{ ergs}$$

if the units be practical, where W is the quantity of work converted into heat, and I, E, and R have the same meanings as before.

So also if we consider any particular portion of a circuit, the resistance of which is r, and of which the difference of potential at the two ends is e, then the quantity of work converted into heat in this portion of the circuit in the unit of time may be expressed by the formulæ

$$w = I^2r = \frac{e^2}{r} = eI \text{ ergs}$$

if the units are absolute

$$= eI \times 10^7 \text{ ergs}$$

if the units are practical.

The heat equivalent of the quantity of work expended in t units of time in a circuit, the resistance of which is R, by a current of strength I, is given by the equations

$$Q = \frac{W}{J} = \frac{I^2Rt}{J} \text{ calories}$$

$$= \frac{I^2Rt}{4\cdot2 \times 10^7} \text{ calories}$$

if the units are absolute

$$= \frac{I^2Rt \times 10^7}{4\cdot2 \times 10^7} = I^2Rt \times \cdot 24 \text{ calories}$$

if the units are practical.

ELECTRIC LIGHT ARITHMETIC.

Example 1. One portion of an electrical circuit was composed of 10 feet of copper wire 7 mils in diameter, and another portion of 5 feet of iron wire 148 mils in diameter. If the resistances of copper and iron are to one another as 112 to 825, compare the heating effects of the current in these two portions of the circuit.

Since the same current flows through both wires the value of I is the same for both, and for their relative resistances we have

$$\frac{R_2}{R_1} = \frac{a_2}{a_1} \times \frac{l_2}{l_1} \times \left(\frac{d_1}{d_2}\right)^2 = \frac{825}{112} \times \frac{5}{10} \times \left(\frac{7}{148}\right)^2$$

$$= \frac{1}{121} \text{ nearly.}$$

Therefore

$$\frac{\text{Heat developed in the copper section}}{\text{,, ,, ,, iron ,,}}$$

$$= \frac{\text{Resistance of copper}}{\text{,, ,, iron}} = \frac{121}{1}.$$

Example 2. The interpolar portion of an electrical circuit consisted of two different wires joined end to end.

One was of copper 3 feet long and 65 mils in diameter,

while the other was of platinum 6 inches long and 35 mils in diameter. The resistances of copper and platinum being to one another as 112 to 1243, compare the quantities of heat developed by the current in these portions of the circuit.

$$\text{Ans. } \frac{\text{Heat in platinum}}{\text{,, ,, copper}} = \frac{638}{100}.$$

Example 3. The internal resistance of a certain dynamo machine is 5 Ohms, that of the lamp is 10 Ohms, and of the leading wires 2 Ohms. Compare the quantities of heat produced in the lamp and in the machine, and find what proportion of the total energy in the circuit will appear in the lamp.

$$\text{Ans. } \frac{\text{Heat in lamp}}{\text{,, ,, machine}} = \frac{10}{5} = \frac{2}{1},$$

$$\frac{\text{Energy in lamp}}{\text{Total energy in circuit}} = \frac{\text{Resistance of lamp}}{\text{Resistance of whole circuit}}$$

$$= \frac{10}{5 + 2 + 10} = \frac{10}{17}.$$

Example 4. Two Swan incandescent lamps each of which has a resistance of 35 Ohms are connected in series by thick wires with a series of 40 Grove cells, the average resistance of each cell being 0·5 Ohm. Compare the quantity of heat developed in each lamp with that developed in each cell of the battery, and find what proportion of the total energy yielded by the battery appears in each lamp.

$$\text{Ans. } \frac{\text{Heat in lamp}}{\text{,, ,, cell}} = \frac{70}{1};$$

$$\frac{\text{Energy in one lamp}}{\text{,, of battery}} = \frac{7}{18}.$$

Example 5. The internal resistance of a voltaic battery is equal to that of 3 metres of a particular wire. Compare the quantities of heat produced both inside and outside the battery when its poles are connected by one metre of this wire with the quantities produced in the same time when they are connected by 37 metres of the same wire.

Taking the resistance of one metre of the given wire as our unit and I_1, I_2 as the intensities of the current in the two cases, we have

$$I_1 = \frac{E}{4} \text{ and } I_2 = \frac{E}{40}; \text{ whence } \frac{I_1}{I_2} = 10,$$

and therefore for the heat developed in the battery in the two cases

$$\frac{Q_1}{Q_2} = \frac{I_1^2 R_1}{I_2^2 R_2} = \left(\frac{I_1}{I_2}\right)^2 = \frac{100}{1} \text{ because } R_1 = R_2.$$

For the heating effects outside the battery

$$\frac{Q_1}{Q_2} = \left(\frac{I_1}{I_2}\right)^2 \frac{R_1}{R_2} = \frac{100}{1} \times \frac{1}{37} = \frac{100}{37}.$$

Example 6. A Grove cell whose electro-motive force is 1·9 Volt and internal resistance 0·4 Ohm has its poles connected (1) by a wire of 3 Ohms, and (2) by a wire of 30 Ohms resistance. Compare the amounts of heat developed in the cell in the two cases.
Ans. As 80 : 1.

Example 7. A dynamo machine the resistance of which is 6 Ohms has its poles connected (1) by a circuit of ·5 Ohm resistance, and (2) by a circuit of 500 Ohms resistance. Compare the quantities of energy which are dissipated as heat in the machine assuming that the driving engine works always at the same horse-power. *Ans.* As 78 : 1 nearly.

Example 8. The interpolar portion of a circuit consisted of a piece of copper and a piece of platinum wire. The platinum was 1 foot long and 28 mils in diameter, while the copper was 20 yards long and 65 mils in diameter. Compare the quantities of heat developed, per running foot, in each of these wires.
Ans. As 60 : 1 nearly.

ELECTRIC LIGHT ARITHMETIC. 35

Example 9. The difference of potential between the ends of the two carbons of a certain Brush arc light is 36 Volts, and a current of 10 Ampères is passing through it. Find the quantity of heat developed per second.

When expressed in the Centimetre-Gramme-Second (C.G.S.) system of units

One Volt = 10^8 C.G.S. units of potential,
One Ohm = 10^9 C.G.S. electro-magnetic units of resistance,
One Ampère = 10^{-1} of the C.G.S. unit of current.

One water-gramme degree Centigrade = $4 \cdot 2 \times 10^7$ ergs.

In the present case we have

$$W = E \times I = 36 \times 10^8 \times 10 \times 10^{-1}$$
$$= 36 \times 10^8 \text{ ergs per second,}$$

$$\therefore Q = \frac{36 \times 10^8}{4 \cdot 2 \times 10^7} = \frac{360}{4 \cdot 2} = 85 \cdot 72 \text{ w.g.d. C}°.$$

Example 10. An Edison incandescent lamp has a resistance of 125 Ohms, and the difference of potential between the ends of the carbon filament is 110 Volts. Find the strength of current flowing through it and the quantity of heat developed in the lamp per second.

Ans. Current = 0·88 Ampère. Heat = 23·05 w.g.d. C°.

Example 11. The house-main wire for supplying 20 Edison lamps, each of which requires a current of 0·8 Ampère is of copper of 86 per cent. conductivity, and 65 mils in diameter. What is the amount of heat developed per second in every foot of this wire?

Following the method of Example 33, page 14, we shall find that the resistance of one foot of this wire is 0·00274 Ohm, and then substituting the numerical values in the equation

$$Q = R \times I^2$$
$$= 2\cdot74 \times 10^{-3} \times 10^9 \times (\cdot 08 \times 20)^2 \text{ ergs per second.}$$
$$= \frac{2\cdot74 \times 10^6 \times (1\cdot6)^2}{4\cdot2 \times 10^7} = 0\cdot167 \text{ w.g d. C}°.$$

Example 12. A current of one Ampère circulates through a wire whose resistance is 0·9536 Ohm. Find the amount of heat developed in this wire in 5 minutes.

Ans. 68 w.g.d. C°.

Example 13. A current of 0·75 Ampère was sent for 5 minutes through a column of mercury whose resistance was 0·47 Ohm. The mass of the mercury was 20·25 grammes and its specific heat 0·0332; find the rise of temperature, assuming that no heat escapes by radiation.

By the method adopted in the previous examples we find that the number of units of heat developed in 5 minutes by the passage of a current of 0·75 Ampère through a resistance of 0·47 Ohm is 18·88 units, and by the question this quantity of heat will raise 20·25 grammes of mercury through $x°$ C. of temperature.

But the amount required for this purpose is

$$= 20\text{·}25 \times 0\text{·}0332 \times x$$
$$= 18\text{·}88 \text{ units by the question.}$$

Hence

$$x = \frac{18\text{·}88}{20\text{·}25 \times 0\text{·}0332} = 28° \text{ C.}$$

Example 14. A coil of very fine wire which had a resistance of 45·64 Ohms was placed in 1000 grammes of water at 0° C. The current from a series of 50 voltaic cells, each of which had an electro-motive force of 1 Volt and a resistance of 6 Ohms, was sent through the coil for 10 minutes. Find the rise of temperature of the water. *Ans.* 1·39° C.

Example 15. A small coil of German silver wire was placed in one of Lenz's alcohol calorimeters, and a steady current of 1·8 Ampère was sent through it. The resistance of the coil was 1·2 Ohm, and it was found that the temperature of the alcohol rose 1° C. in 23 seconds. Had the resistance of the coil been equal to 1 Ohm while the current strength was the same, what would have been the time required?

Ans. 27·6 seconds.

Example 16. A wire of pure copper, 0·165 centimetre in diameter, has a current of 10 Ampères flowing through it.[1] What will be the limiting temperature of the wire?

The specific resistance of pure copper per cubic centimetre is 1·65 Microhm, and therefore the resistance of one metre of the above wire will be $7·8 \times 10^{-3}$ Ohm. Therefore the quantity of heat developed per second in one metre of this wire is

$$= \frac{7·8 \times 10^6}{4·2 \times 10^7} = ·1857 \text{ thermal unit.}$$

The surface of a length of one metre of this wire is

$$= 100 \times \pi \times 0·165 = 51·8 \text{ square centimetres.}$$

Hence the development of heat in this wire is at the rate of $\frac{·1857}{51·8}$ or 0·00359 of a thermal unit per square centimetre of surface per second. But the rate of loss of heat by radiation and convection from an unpolished surface of copper (or other solid material) is about 1-4000th per square centimetre per degree of excess of temperature above that of the surrounding medium.

---

[1] This assumption is true only in certain exceptional cases.

ELECTRIC LIGHT ARITHMETIC. 39

It follows therefore that the temperature of this wire will go on rising until it has reached such a value that the loss per second by radiation and convection is equal to 0·00359 of a thermal unit. Hence if $x°$ be this temperature

$$\frac{x}{4000} = ·00359 \text{ and } \therefore x = 14°·36 \text{ C.}$$

Hence, if this wire be freely exposed to the air, its limiting temperature will be 14°·36 C. above that of the surrounding air.

Example 17. Taking account of the loss of heat by radiation, calculate what would be the rise of temperature of the wire referred to in Example 11.

Ans. 42°·3 C.

Example 18. A short piece of lead wire is included as a *safety catch* in a circuit, and it is required to find what its diameter should be so that a current of 7·2 Ampères may just fuse it, assuming that the specific resistance of lead per cubic centimetre is 19·85 Microhms, and that the melting point of lead is 335° C.

Let x = diameter of the wire in centimetres. The resistance of a lead wire one centimetre long and x centimetres in diameter is

$$= 19·85 \times 10^{-6} \times \frac{4}{\pi x^2} \text{ Ohms,}$$

and therefore the quantity of heat developed in one second in one centimetre of this wire

$$= \frac{(\cdot 72)^2 \times 19\cdot 85 \times 10^3 \times 4}{\pi x^2 \times 4\cdot 2 \times 10^7} \text{ w.g.d. C}°.$$

But the area of the surface of one centimetre of this wire is πx square centimetres, and therefore the development of heat per square centimetre of surface is at the rate of

$$\frac{(\cdot 72)^2 \times 19\cdot 85 \times 10^3 \times 4}{\pi^2 x^3 \times 4\cdot 2 \times 10^7} \text{ thermal units per second.}$$

Now while the current is flowing the wire will go on rising in temperature, but the rate of increase will be slower and slower as the temperature approaches the limiting one at which the gain of heat per second is just balanced by the loss by radiation and convection, and if this limiting temperature be equal to or higher than that at which lead fuses then the wire will melt and the current will be interrupted. If therefore the diameter of the wire is to be such that the wire shall *just* melt with the above current we have the equation

$$\frac{335}{4000} = \frac{(\cdot 72)^2 \times 19\cdot 85 \times 10^3 \times 4}{\pi^2 x^3 \times 4\cdot 2 \times 10^7}$$

wherce

$$x^3 = \frac{4000 \times (\cdot 72)^2 \times 19\cdot 85 \times 10^3 \times 4}{335 \times \pi^2 \times 4\cdot 2 \times 10^7}$$

ELECTRIC LIGHT ARITHMETIC. 41

and

∴ $x =$ ·106 centimetre, or about No. 19 B.W.G.

For the sake of simplicity it has been assumed in this solution that the resistance of the lead does not vary with the temperature, and also that the rate of loss of heat by radiation follows the same law at comparatively high temperatures. The student should also notice that the value of the *specific heat* of lead only affects the *time* which would elapse from the commencement of the flow before the wire fuses, and does not affect the practical question as to whether or not fusion will ultimately take place.

Example 19. A leaden *safety catch* is to be inserted in a circuit which will fuse if the current strength exceeds 20 Ampères. What must be its diameter?

N.B.—A *safety catch* is a short piece of wire whose thickness and material are so chosen that if it be inserted in a circuit it will fuse and so interrupt the circuit if the strength of the circuit exceeds a given value. In the present case we shall find by the method adopted in Example 18 that $x =$ ·209 centimetre, or the wire is about No. 14 B.W.G.

Example 20. A copper wire is to be inserted in a circuit as a safety catch for a current of 500 Ampères. Taking the melting point of copper at 1050° C. and its specific resistance per cubic centimetre equal to 1·652 Microhm; find what must be its diameter.

Ans. ·53 centimetre, or about No. 6 B.W.G.

Example 21. If a piece of zinc rod is to be used as a safety catch for a current of 500 Ampères, find what must be its diameter, having given that the melting point of zinc is 422° C. and its specific resistance per cubic centimetre is 5·689 Microhms.

Ans. 1·091 centimetre, or about No. ooo B.W.G.

Example 22. A piece of platinum wire 0·55 millimetre in diameter is inserted in a circuit. If the resistance of one metre of platinum wire one millimetre in diameter be 0·1166 Ohm and the melting point of platinum be 1700° C., calculate the strength of current which will be required to fuse the above wire.

The resistance of a length of one centimetre of this wire is

$$= \frac{·1166}{100} \times \left(\frac{1}{·55}\right)^2 \text{ Ohm.}$$

and therefore the quantity of heat developed per second by a current of x Ampères is

$$= \frac{\left(\frac{x}{10}\right)^2 \times ·1166 \times 10^9}{100 \times (·55)^2 \times 4·2 \times 10^7} \text{ w.g.d C°.}$$

But the surface of one centimetre of this wire $= \pi \times {}^{\cdot}055$ square centimetre, and therefore the heat developed per second per square centimetre of surface is

$$= \frac{x^3 \times {}^{\cdot}1166}{100 \times ({}^{\cdot}55)^2 \times \pi \times {}^{\cdot}055 \times 4{}^{\cdot}2} \text{ w.g.d. C}^{\circ}.$$

and if the limiting temperature be that of the melting point of platinum this expression must be equal to $\frac{1700}{4000}$ and therefore

$$x^2 = \frac{17 \times 100 \times ({}^{\cdot}55)^2 \times 4{}^{\cdot}2 \times \pi \times {}^{\cdot}055}{40 \times {}^{\cdot}1166}$$

and $\therefore x = 8{}^{\cdot}9$ Ampères nearly.

Example 23. A lead wire which is 0·81 millimetre in diameter is to be used as a safety catch. What strength of current will it just bear without fusing?

Ans. 4·76 Ampères.

Example 24. Twenty of Edison's incandescent lamps, each of which requires a current of 0·8 Ampère, are to be supplied from one main wire. If a margin of 50 per cent. excess of current be allowed, find the thickness of a lead safety catch which should be inserted in the supply wire.

The fusing current is in this case one of 24 Ampères

and then working the question out as before we find the diameter required = 0·236 centimetre.

Example 25. A piece of lead wire of the same diameter as the copper conductor is inserted in an electrical circuit and a strong current is sent through the circuit until the lead wire fuses. Assuming that there is no loss of heat by radiation, find the temperature of the copper conductor at the moment when this occurs.

If we take the specific resistance of lead per cubic centimetre to be 19·85 Microhms, and that of copper 1·65 Microhm, the ratio of the resistances of similar wires of lead and copper will be very nearly as 12 : 1, and therefore the quantities of heat developed in equal lengths by the same current in one second will be as 12 to 1.

But the specific gravity of lead is 11·38, and that of copper is 8·89, while the specific heat of lead is 0·0314, and that of copper is 0·0948.
Therefore

$$\frac{\text{Heat required to raise a given volume of lead 1°C.}}{\text{,, ,, ,, same ,, copper ,,}} = \frac{11\cdot38 \times 0\cdot0314}{8\cdot89 \times 0\cdot0948},$$

and consequently while the temperature of the copper wire is being raised 1 °C. the temperature of the same length of lead wire of the same diameter through which the same current is flowing will be raised

$$\frac{8\cdot 89 \times 0\cdot 0948}{11\cdot 38 \times 0\cdot 0314} \times \frac{12}{1} = 28^{\circ}\cdot 3 \text{ C.}$$

But the melting point of lead is 335° and therefore by the time the lead is heated to 335° C. the copper will have had its temperature raised by

$$\frac{335}{28\cdot 3} = 11^{\circ}\cdot 8 \text{ C.}$$

The student will understand that this result is only approximately true, because no notice has been taken of the change of resistance with increasing temperature, nor of the loss of heat by radiation and convection.

WORK UTILISED IN A SIMPLE CIRCUIT.

Example 1. The difference of potential between the electrodes of a certain Swan lamp was 31·1 Volts, and the current flowing through it was 1·22 Ampère. Calculate the energy, in horse-power, absorbed by the lamp, and also the amount of heat developed.

By Joule's formula we have

$$W = I^2 R = I \times E,$$

and expressing the values of the Volt and Ampère in absolute C.G.S. measure we have in the present case

$$W = \frac{1\cdot22}{10} \times 31\cdot1 \times 10^8 \text{ ergs}$$

$$= 3\cdot7942 \times 10^8 \text{ ergs per second.}$$

But one horse-power is equivalent to $7\cdot46 \times 10^9$ ergs, and therefore the energy absorbed by the lamp per second

$$= \frac{3\cdot7942 \times 10^8}{7\cdot46 \times 10^9} = \frac{1}{20}\text{th of a H.P.}$$

Again, one water-gramme-degree Centigrade is equivalent to $4\cdot2 \times 10^7$ ergs, and therefore the amount of heat developed in this lamp per second is

$$= \frac{3\cdot7942 \times 10^8}{4\cdot2 \times 10^7} = 9 \text{ w.g.d. C}°.$$

Example 2. The strength of current flowing through a certain lamp was found to be 2·42 Ampères and the electro-motive force between the terminals was 57 Volts. What amount of horse-power was being expended in the lamp?   *Ans.* 0·185 H.P.

Example 3. What horse-power is required to drive a current of 4·16 Ampères through a resistance of 21·4 Ohms?

By Joule's formula we have

$$W = I^2 \times R = (.416)^2 \times 21.4 \times 10^9 \text{ ergs}$$

$$= \frac{(.416)^2 \times 21.4 \times 10^9}{7.46 \times 10^9} \text{ H.P}$$

$$= \frac{1}{2} \text{ H.P. nearly.}$$

Example 4. The resistances of three incandescent lamps which are placed in series are 60, 32, and 50 Ohms respectively. What horse-power will be required to send a current of one Ampère through them, neglecting the resistance of the connecting wires?
Ans. 0·19 H.P.

Example 5. A certain dynamo machine has an internal resistance of 3 Ohms, and that of the lamps and leading wires is 10 Ohms. What horse-power is expended in driving a current of 15 Ampères through this circuit? *Ans.* 3·9 H.P.

Example 6. Find the amount of horse-power required to maintain a current of 10 Ampères through 8 arc lamps, each of which has a resistance of 6 Ohms, the resistance of the leading wires being 8 per cent. of that of the lamps. *Ans.* 7 H.P. nearly.

Example 7. Suppose that we have 16 Brush lamps in series, that each lamp has a resistance of 4·4 Ohms, and that the resistance of the rest of the circuit is 12·65 Ohms. If 60 per cent. of the total energy expended appears in the lamps, find the amount of horse-power required to maintain a current of 10 Ampères. *Ans.* 15·73 H.P.

Example 8. A certain size of Swan lamp has, when hot, a resistance of 30 Ohms. There are 30 of these lamps in series and the resistance of the remainder of the circuit is 15 Ohms. What horse-power will be required to maintain a current of 1·5 Ampère through this series of lamps? *Ans.* 2·76 H.P.

Example 9. An electric circuit is composed of 5 arc lamps, each of which has a resistance of 2·5 Ohms, the armature of the machine which has a resistance of 2 Ohms, and the leading wires which have a resistance of 1·5 Ohm. Find what proportion of the total energy expended in the circuit appears in the lamps.

The total resistance of the circuit = 16 Ohms, and the ratio of the work expended in any part of the circuit to the total work is equal to the ratio of the resistance of that part to the resistance of the whole circuit. Hence in the present case the proportion of the total energy which is developed in the lamps

$$= \frac{12 \cdot 5}{16} = \cdot 78 \text{ or } 78 \text{ per cent.}$$

Example 10. If the internal resistance of a dynamo machine be 3 Ohms and that of the external circuit 18 Ohms: what is the ratio of the external or useful work of the current to the total work expended?
Ans. 6-7ths.

Example 11. The resistances of two incandescent lamps arranged in series are 50 and 100 Ohms respectively, and the total resistance of the circuit is 200 Ohms. What fraction of the total work is expended in each of these lamps? *Ans.* ¼ and ½.

Example 12. In an experiment with a certain dynamo machine the resistance of the armature was 0·016 Ohm, and that of the external circuit 0·757 Ohm The power required to turn the armature at a certain rate in the magnetic field was 7·604 H.P. and this produced a current of 83·7 Ampères. Calculate the *duty* of the generator and also its *commercial efficiency*.

The total resistance of the circuit = 0·757 + 0·016 = 0·773 Ohm, and therefore by the method of Example 3 we find that the electrical energy in the current

$$= \frac{(8·37)^2 \times 0·773}{7·46} = 7·259 \text{ H.P.}$$

The *duty* of the generator is the ratio of the total electrical energy developed to the amount of energy expended in turning the armature in the magnetic field. In the present case it is

$$= \frac{7\cdot259}{7\cdot604} = 0\cdot955.$$

The *commercial efficiency* of the generator is the ratio of the amount of electrical energy which appears in the external circuit to the total energy which is expended in driving the machine. In the present case the commercial efficiency

$$= \frac{757 \times 7\cdot259}{773 \times 7\cdot604} = 0\cdot935.$$

Example 13. The internal resistance of a certain dynamo machine was 1·5 Ohm, and the external resistance 2·7 Ohms. An expenditure of 3·87 H.P. in turning the armature developed a current of 25·5 Ampères. Calculate from these data the *duty* of the generator and also its *commercial efficiency*.

Ans. Duty = 0·95 ; Commercial efficiency = 0·61.

Example 14. In an experiment with a Brush machine the work expended in driving the armature was 15·3 H.P., the internal resistance was 10·5 Ohms, the external 73 Ohms, and the strength of the current was 10 Ampères. Find the *commercial efficiency* of this machine. *Ans.* 0·64.

ELECTRIC LIGHT ARITHMETIC. 51

Example 15. The external resistance of an electrical circuit being 0·923 Ohm, and the internal resistance of the Gramme machine 1·077 Ohm, an expenditure of 4 H.P. in driving the machine produced a current of 32 Ampères. What was the *commercial efficiency* of this machine? *Ans.* 0·32 nearly.

Example 16. What was the *duty* of the machine alluded to in the last example? *Ans.* 0·69 nearly.

Example 17. The resistance of a certain Gramme machine is 1·67 Ohm, and that of the leading wires 0·03 Ohm, while that of the lamp is 2 Ohms. If 3·5 horse-power is expended in driving the armature, and the *duty* of the machine is 0·72; find the quantity of energy which will be expended in the lamp.

The electrical energy in the circuit = 3·5 × 0·72 = 2·52 H.P.
Total resistance of circuit = 1·67 + ·03 + 2 = 3·7 Ohms.
Resistance of lamp = 2 Ohms.
Hence the quantity of energy expended in the lamp

$$= \frac{2}{3 \cdot 7} \times 2 \cdot 52$$
$$= 1 \cdot 36 \text{ H.P.}$$

If we wish to know the value in Ampères of the current flowing through the lamp, we can calculate it from the formula

$$W = I^2R \text{ whence } I = \sqrt{\frac{W}{R}}$$

and in the present case

$W = 1.36 \times 7.46 \times 10^9$ ergs.
$R = 2 \times 10^9$ electro-magnetic C.G.S. units of resistance.

therefore

$$I = \sqrt{\frac{1.36 \times 7.46}{2}} = 2.25 \text{ C.G.S units of current,}$$
$$= 22.5 \text{ Ampères.}$$

Example 18. Three arc lamps, each of which had a resistance of 1·5 Ohm, were joined up in series with a dynamo machine whose resistance was 2·8 Ohms. The resistance of the leading wires was 0·22 Ohm, and the *duty* of the machine 0·95. Find the quantity of energy developed in each lamp when 5·45 H.P. was expended in driving the armature.
Ans. 1·03 H.P.

Example 19. What was the strength of current flowing through each lamp? *Ans.* 22·6 Ampères.

Example 20. A certain Brush machine had an internal resistance of 5·06 Ohms, and it was connected up by wires of 1·3 Ohm resistance with a series of 16 lamps, each of which had a resistance of 3·8 Ohms.

ELECTRIC LIGHT ARITHMETIC. 53

The *duty* of the machine was 0·73. What was the amount of energy developed in each lamp when 14 H.P. was being expended in driving the armature?

Ans. 0·58 H.P.

Example 21. What was the strength of current, in Ampères, flowing through each lamp?

Ans. 10·67 Ampères.

Example 22. The internal resistance of a Gramme machine provided with a permanent steel magnet of constant strength was 0·4 Ohm. When the terminals were insulated the power required to drive the bobbin at a certain rate was 0·01 H.P. When the terminals were connected by a wire of 3·6 Ohms resistance, the power required to drive the bobbin at the same rate as before was 0·1 H.P. What power would be required to drive the bobbin at the same rate if the terminals were connected by a wire of 0·2 Ohm resistance? If this wire be immersed in one pound of cold water through how many degrees per minute would its temperature be raised, the mass of the wire being negligible?

Since the machine is provided with a permanent magnet of constant strength, and since the rate of revolution of the armature is constant, it follows that the electro-motive force of the machine is constant.

But by Joule's equation

$$W = I \times E = \frac{E^2}{R} = \frac{a}{R} \text{ where } a \text{ is a constant.}$$

Now by the question 0·01 H.P. is used in overcoming friction, and therefore when the total energy expended is 0·1 H.P. and the resistance $R = 0·4 + 3·6 = 4$ Ohms, we have

$$\frac{a}{4} = 0·1 - 0·01 = 0·09 \text{ H.P.} ; \therefore a = ·36.$$

Let x represent the horse-power required in the second case, then since

$$W = \frac{·36}{R}$$

we get

$$W = \frac{·36}{·4 + ·2} = \frac{3·6}{6} = ·6 \text{ H.P.}$$

and therefore

$$x = ·6 + ·01 = ·61 \text{ H.P.}$$

The energy expended in the wire

= 1-3rd of the total electrical energy in circuit,
= 1-3rd × ·61 H.P. = 0·2 H.P ;

and this expressed in water-pound degrees Fahrenheit

$$= \frac{0·2 \times 33000}{772} = 8°·55 \text{ F}$$

ELECTRIC LIGHT ARITHMETIC. 55

Example 23. The resistance of the armature of one of Edison's large dynamo machines is 0·008 Ohm. If a current of 900 Ampères be supplied by it, find the quantity of heat developed per second in the armature *Ans.* 1543 w.g.d. C°.

Example 24. The difference of potential between the two terminals of a certain arc lamp was 37·7 Volts, and a current of 25 Ampères was flowing through it. Express, in "Watts," the power consumed in the lamp.

We have

$$W = I \times E = 25 \times 37.7 = 942.5 \text{ Watts.}$$

Note.—In his Presidential Address to the British Association, Dr. C. W. Siemens has suggested the adoption for the electro-magnetic unit of power, of the amount of power which is conveyed by a current of one Ampère through a difference of potential of one Volt. This unit is to be called a Watt, in honour of the celebrated mechanician James Watt.

Example 25. Express in horse-power the power consumed in the above-mentioned lamp.

$$\text{One horse-power} = 7.46 \times 10^9 \text{ ergs.}$$
$$\text{One Watt} = 10^{-1} \times 10^8 = 10^7 \text{ ergs.}$$
$$\therefore 942.5 \text{ Watts} = \frac{942.5}{746} = 1.26 \text{ H.P.}$$

56   ELECTRIC LIGHT ARITHMETIC.

Example 26. A current of 9·2 Ampères is flowing through an arc lamp with a difference of potential between the terminals of 46·6 Volts. How many Watts are consumed in this lamp, and what will be the number representing this amount in horse-power?
Ans. 428·7 Watts; 0·58 h.p.

Example 27. Two incandescent lamps, each of which has a resistance of 73 Ohms, are joined in series by thick wire. What power will be expended in driving a current of 2·5 Ampères through them?
Ans. $W = I^2 \times R = (2·5)^2 \times 146 = 912·5$ Watts.

Example 28. The internal resistance of a certain dynamo machine is 2·5 Ohms, and that of the lamps and leading wires 11 Ohms. What power is expended in driving a current of 16 Ampères through this circuit? *Ans.* 3456 Watts.

Example 29. The current through a certain glow lamp is 1·07 Ampère, and at its terminals there is a steady difference of potential of 50 Volts while the candle-power is 14. What is the illuminating power of this lamp in candles per horse-power?

Since expenditure $= \dfrac{1·07 \times 50}{746}$ h.p.

ELECTRIC LIGHT ARITHMETIC. 57

∴ One candle of illumination requires an expenditure of $\frac{1\cdot07 \times 50}{746 \times 14}$ h.p.

∴ One h.p. will produce $\frac{746 \times 14}{1\cdot07 \times 50}$, or 195·2 candles light, nearly.

Example 30. If the candle-power were 23 for a current of 1·2 Ampères, and a potential difference at the terminals of 55 Volts, what would be the light-producing power of the lamp per horse-power?
Ans. 260 C.P. nearly.

Example 31. In the case of another lamp the illumination was found to be 16 C.P. for 50 Volts electrical pressure, and a current of 1·05 Ampères. What was the illuminating power of this lamp per horse-power?   *Ans.* 227·4 C.P. nearly.

COMPOUND ELECTRICAL CIRCUITS.

Example 1. Two points in an electrical circuit are connected by two incandescent lamps, whose resistances are 31 and 37 Ohms respectively. Find the resultant resistance of the circuit between these two points.

The conductivity of the two branches is

$$\frac{1}{31} + \frac{1}{37} = \frac{68}{31 \times 37},$$

and therefore their resultant resistance, which is the reciprocal of their conductivity, is

$$= \frac{31 \times 37}{68} = 16\cdot87 \text{ Ohms.}$$

Example 2. Three conducting wires, whose resistances are 5, 7, and 9 Ohms, are joined together in multiple arc. Find the resultant resistance of this compound conductor. *Ans.* 2·2 Ohms.

Example 3. Four wires are joined together in multiple arc, their resistances being 5·5, 18, 3·7, and 2·9 Ohms respectively. Find the resultant resistance of the compound conductor thus formed.
Ans. 1·17 Ohm.

Example 4. An iron wire 20 metres long and 0·2 centimetre in diameter, a copper wire 105 metres long and 0·15 centimetre in diameter, and a German silver wire 0·6 metre long and 0·04 centimetre in diameter were joined together in multiple arc. Find the resultant resistance of the compound conductor thus formed, having given that

MICROHMS
Specific resistance of iron per cubic centimetre = 9·825
,, ,, copper ,, ,, = 1·652
,, ,, German silver ,, ,, = 21·170
Ans. 0·277 Ohm.

ELECTRIC LIGHT ARITHMETIC. 59

Example 5. The poles of a dynamo machine are connected in multiple arc by 20 incandescent lamps, each of which has a resistance of 80 Ohms. The internal resistance of the machine being 0·25 Ohm, what must be its electro-motive force so that a current of 1·2 Ampère may flow through each lamp?

The resistance of the interpolar circuit $= \frac{80}{20} = 4$ Ohms.

,, ,, ,, machine $= 0\cdot25$ Ohm.

∴ total resistance of the circuit $= 4\cdot25$ Ohms.

The current flowing through the machine

$$= 20 \times 1\cdot2 = 24 \text{ Ampères,}$$

and therefore for the value of the electro-motive force in circuit we have by Ohm's law

$$E = I \times R = 24 \times 4\cdot25 = 102 \text{ Volts.}$$

Example 6. Under the circumstances described in the last Example, what would be the electro-motive force in the outer circuit between the terminals of the machine?

This can readily be calculated from the known values of the resistance of any lamp and the strength of current flowing through it.

By Ohm's law

$$E = I \times R = 1\cdot2 \times 80 = 96 \text{ Volts.}$$

Example 7. A thousand incandescent lamps are joined up in simple multiple arc to the poles of a dynamo machine whose resistance is 0·004 Ohm. If the average resistance of each lamp is 120 Ohms, and it requires a current of 0·9 Ampère to vitalise it, what must be the electro-motive force of the machine?

Ans. 111·6 Volts.

Example 8. The internal resistance of a dynamo machine is 0·008 Ohm, and it is employed to drive a current of 0·8 Ampère through each of 900 incandescent lamps arranged in simple multiple arc. The resistance of each lamp when hot being 130 Ohms, what must be the electro-motive force of the machine?

Ans. 109·4 Volts.

Example 9. The electro-motive force of a certain dynamo machine is 45 Volts, and its resistance is 0·01 Ohm. The lamps to be used with it have a hot resistance of 35 Ohms. What number may be arranged in simple multiple arc so that a current of 1·2 Ampère may flow through each of them?

If $x =$ number of lamps, then the resultant resistance of the external circuit is $\frac{35}{x}$ Ohms, and the total resistance of the circuit is $\left(\frac{35}{x} + ·01\right)$ Ohms. Also the current flowing through the dynamo is $x \times 1·2$ Ampère, and using Ohm's equation

$$E = I \times R,$$

ELECTRIC LIGHT ARITHMETIC. 61

we have

$$45 = x \times 1\cdot 2 \left(\frac{35}{x} + \cdot 01\right)$$
$$= 1\cdot 2 \times 35 + x \times \cdot 012$$

and $\therefore x = 250$ lamps.

Example 10. Find the number of lamps which can be driven in simple multiple arc by a dynamo machine whose resistance is 0·032 Ohm, and its electro-motive force 55 Volts, if each lamp has a hot resistance of 28 Ohms and requires a current of 1·6 Ampère.

Ans. 199 lamps.

Example 11. The internal resistance of a certain dynamo machine is 1 Ohm, and its electro-motive force is 484 Volts when the interpolar portion of the circuit consists of 200 incandescent lamps arranged in 20 series of 10 each. If the resistance of each lamp be 30 Ohms, what is the strength of current flowing through it?

The resistance of each series
 of lamps . . . = 30 × 10 = 300 Ohms.
The resultant resistance of
 the 20 series . . . = $\frac{300}{20}$ = 15 Ohms.
Total resistance of the
 whole circuit . . = 15 + 1 = 16 Ohms.

Hence the current flowing through the machine

$$= \frac{484}{16} = 30\cdot 25 \text{ Ampères,}$$

and therefore the strength of the current through each one of the 20 series is

$$= \frac{30\cdot 25}{20} = 1\cdot 51 \text{ Ampère.}$$

Example 12. Thirty-five incandescent lamps, each of which has a hot resistance of 35 Ohms, are arranged in 5 series of 7 each, and are then connected in multiple arc with the poles of a dynamo machine whose internal resistance is 2·5 Ohms and its electro-motive force 309 Volts. What is the strength of current through each lamp? *Ans.* 1·2 Ampère.

Example 13. The resistance of a certain dynamo machine is 2 Ohms, and it is required to arrange 80 incandescent lamps, each of which has a hot resistance of 50 Ohms, in such a way that the resistance in the external circuit shall be 20 times that of the machine. How is this to be done?

Let x = number of series. Then $\frac{80}{x}$ = number of lamps in each series, and

$$\frac{80 \times 50}{x^2} = \text{the resistance of the external circuit,}$$

$$= 40 \text{ Ohms, by the conditions of the problem.}$$

ELECTRIC LIGHT ARITHMETIC. 63

Hence

$$x^2 = \frac{80 \times 50}{40} = 100; \therefore x = 10 \text{ rows}$$

Example 14. A set of 120 incandescent lamps, each of 30·7 Ohms resistance, are to be arranged in such a way that their resultant resistance shall be about 16 times as great as that of the dynamo machine, whose resistance is 1·6 Ohm. How is this to be done?

Ans. 12 series of 10 each.

Example 15. A certain kind of incandescent lamp requires a current of 2·5 Ampères with an electro-motive force of 73 Volts to work it. There are 60 of these lamps, arranged in 20 series of 3, and the resistance of the machine is 0·27 Ohm. What must be its electro-motive force?

The resistance of each lamp $= \frac{73}{2\cdot 5} = 29\cdot 2$ Ohms

„ „ external circuit $= \frac{29\cdot 2 \times 3}{20} = 4\cdot 38$ Ohms.

Total resistance of circuit $= 4\cdot 38 + \cdot 27 = 4\cdot 65$ Ohms.

But since there are 20 series, and each requires a current of 2·5 Ampères, the current through the dynamo machine $= 20 \times 2\cdot 5 = 50$ Ampères, and therefore the requisite electro-motive force of the machine

$= 50 \times 4\cdot 65 = 232\cdot 5$ Volts.

64   ELECTRIC LIGHT ARITHMETIC.

Example 16. The current required for a certain incandescent lamp is 1·2 Ampère, with an electro-motive force of 63·6 Volts. A hundred of these lamps are joined up in 50 series of 2 each to the poles of a dynamo machine whose internal resistance is 0·14 Ohm. What must be its electro-motive force?

Ans. 135·6 Volts.

Example 17. Forty-five incandescent lamps, each of which has a resistance of 98 Ohms, are arranged in 15 series, and a current of 0·67 Ampère is sent through each. If the resistance of the dynamo be 1·23 Ohm, what must be its electro-motive force?

Ans. 209·3 Volts.

Example 18. Find the number of Daniell cells which would be required to construct a voltaic battery having the same electro-motive force and internal resistance as a certain Brush machine whose electro-motive force is 839 Volts, and its resistance is 10·55 Ohms, having given that the electro-motive force of a Daniell cell is one Volt, and its resistance equal to 5 Ohms.

Let x = number of cells in one series, and y the number of series; then, since the EMF of a Daniell = one Volt, we have $x = 839$.

The resistance of the battery $= \dfrac{839 \times 5}{y} = 10\cdot 55$
by the question;

$$\therefore y = \dfrac{839 \times 5}{10\cdot 55} = 397\cdot 63,$$

also total number of cells required
= 397·63 × 839 = 333612 cells.

Example 19. A certain dynamo machine has an electro-motive force between its terminals of 110 Volts, and the resistance of the whole circuit is 0·11 Ohm. If a Grove's cell has an electro-motive force of 1·97 Volt, and on the average a resistance of 0·37 Ohm, find the number of such cells which would have to be grouped together so as to produce a current equal to that of the above machine through a short thick wire.
Ans. 10487 cells.

Example 20. A Gramme machine, whose resistance is 0·1 Ohm, is employed to send a current through 16 incandescent lamps, each of which has a resistance of 25 Ohms. The electro-motive force of the machine is proportional to the speed at which it is driven. If the driving engine work always at the same horse-power, and if the armature of the Gramme make 1200 revolutions per minute when all the lamps are in series, find the number of revolutions it will make when all the

F

lamps are—(1) in multiple arc, (2) arranged in 4 series of 4.

Let E represent the electro-motive force of the machine, and I the strength of the current flowing through the machine. Then since the amount of energy converted into electricity is constant we have

$$E \times I = \text{a constant quantity,}$$

and $\therefore E = \dfrac{m}{I}$ where m is some constant.

Again by Ohm's law we have

$$I = \frac{E}{R}; \therefore m = \frac{E^2}{R} \quad \ldots \quad (1).$$

But $R = B + L$, where L = resultant resistance of the lamps and B = the resistance of the machine.

When all the lamps are in series, $L = 16 \times 25 = 400$ Ohms;

$$\therefore R = 400 \cdot 1 \text{ Ohms.}$$

If x = number of revolutions per minute, then $E = nx$ where n is a constant; and substituting this value for E in equation (1) we get

$$m = \frac{n^2 x^2}{R}; \therefore x^2 = R \times q \text{ where } q \text{ is some constant.}$$

But when $R = 400\cdot1$, $x = 1200$;

$$\therefore q = \frac{(1200)^2}{400\cdot1} = 3599.$$

When all the lamps are in multiple arc,

$$R = 0\cdot1 + 1\cdot563 = 1\cdot663 \text{ Ohms};$$

$\therefore x^2 = 3599 \times 1\cdot663$, whence $x = 77$ revolutions.

Again, when the lamps are arranged in 4 series of 4, $R = 25\cdot1$, and then

$$x^2 = 3599 \times 25\cdot1 \; ; \; \therefore x \, 301 \text{ revolutions.}$$

Example 21. Sixty accumulators are arranged in series to feed a number of glow-lamps arranged in parallel. Each accumulator has an internal resistance of $\cdot005$ Ohm and a discharge electromotive force of 2 Volts, while each lamp requires a current of $\frac{1}{2}$ an Ampère and 111 volts potential difference between its terminals. How many lamps can be thus fed?

Let $x =$ number of lamps, then the current through the accumulators is $\frac{x}{2}$ Ampères.

Since the total fall of potential is equal to the sum of the fall of potential in the accumulators and of that in the lamp circuit, we have

$$\frac{x}{2} \times \cdot 005 \times 60 + 111 = 60 \times 2$$

$$\therefore x = 60 \text{ lamps.}$$

Example 22. How many lamps could be fed by 30 of the above accumulators, if each lamp required one Ampère of current and 50 volts of electrical pressure?
Ans. 66 lamps.

Example 23. If each lamp required 90 Volts and 2·25 Ampères of current, how many lamps could be run in parallel by 47 such accumulators in series?
Ans. 7·6 lamps nearly.
That is, 8 lamps all a little *below* normal brightness,
or 7 „ „ *above* „ „

Example 24. How many of these lamps could be run in parallel with 48 of these accumulators in series?
Ans. 11 lamps all a little *above* normal brightness,
12 „ „ *below* „ „

Example 25. How many lamps similar to those described in Example 21 could have been run in parallel by 61 accumulators of the kind mentioned in that example?
Ans. 72 lamps all a little *above* normal brightness.

Example 26. Each of 30 glow-lamps arranged in

ELECTRIC LIGHT ARITHMETIC. 69

parallel requires 100 Volts and ·75 of one Ampère. How many accumulators arranged in series will be required to run them if each accumulator has an internal resistance of 0·0059 Ohm and an electromotive force on discharge of 1·8 Volts?

Let x be the number of accumulators required; then, as in the solution of Example 21, we have

$$x \times \cdot 0059 \times 22 \cdot 5 + 100 = x \times 1 \cdot 8;$$
$$\therefore x = 60 \text{ very nearly.}$$

Example 27. If each lamp required not less than 50 Volts and 1½ Ampères of current, and there were 20 lamps in parallel, how many of the above accumulators would be required?

Ans. 31 very nearly.

DISTRIBUTION OF ENERGY IN A COMPOUND CIRCUIT.

Example 1. Two coils of platinum wire, whose resistances are as 5 : 3, are placed in vessels of water forming calorimeters, and are connected in series in an electric circuit, when it is found that the quantities of heat produced in 10 minutes are 72·22 and 43·33 units respectively. The coils are then placed in a multiple arc, and the quantities of heat produced in 10 minutes are then found to be 2·6 and 4·33 units respectively.

Deduce from these experiments Joule's law as to the heating effect of an electric current.

(1) When the coils are in series.

Let H_1 represent the heat developed in 1st coil, and R_1 its resistance.

Let H_2 represent the heat developed in 2nd coil, and R_2 its resistance.

Then we have

$$\frac{H_1}{H_2} = \frac{72\cdot22}{43\cdot33} \text{ and } \frac{R_1}{R_2} = \frac{5}{3},$$

whence

$$\frac{R_1 H_2}{R_2 H_1} = \frac{5 \times 43\cdot33}{3 \times 72\cdot22} = 1;$$

$$\therefore \frac{H_2}{H_1} = \frac{R_2}{R_1} \quad \ldots \quad (a).$$

(2) When the coils are arranged in multiple arc.

In this case the current will divide itself between the two coils in the inverse ratio of their resistances, and therefore the current in the first coil will be 3-8ths, and that through the second coil 5-8ths of the main current.

Let I_1 and I_2 represent the strengths of the currents in the two coils, then we shall have

$$\frac{I_1^2 \times R_1}{I_2^2 \times R_2} \times \frac{H_2}{H_1} = \frac{9 \times 5 \times 4\cdot 33}{25 \times 3 \times 2\cdot 6}$$

$$= \frac{12\cdot 99}{13\cdot 00} = 1, \text{ very nearly }.$$

Therefore

$$\frac{H_2}{H_1} = \frac{I_2^2 \times R_2}{I_1^2 \times R_1} \quad \ldots \quad \ldots \quad (\beta)$$

and from (α) and (β) we get Joule's equation—

$H = KRI^2$ where K is some constant.

Example 2. A current of 2·4 Ampères is sent through two lamps in series, each of which has a resistance of 60 Ohms. The lamps are then placed in multiple arc and the same current is divided between them. Compare the quantities of heat developed in each lamp in the two cases. *Ans.* As 4 : 1.

Example 3. The resistance of a certain dynamo machine is 2 Ohms, and when it is making 820 revolutions a minute it can drive a current of 2·5 Ampères through a single extrapolar resistance of 30 Ohms. The extrapolar part is then made up of 5 incandescent lamps, placed in multiple arc, and each having a resistance of 150 Ohms. Assuming that the electromotive force is proportional to the velocity of rotation, find the rate at which the armature must be made to turn so as to send a current of 1 Ampère through each of these lamps.

Under the conditions stated in the first part of the problem, the electro-motive is found directly by Ohm's law—

$$E_1 = I_1 \times R_1 = 2\cdot5 \times (2 + 30) = 80 \text{ Volts.}$$

In the second arrangement the resultant resistance of the circuit is

$$R_2 = 2 + \frac{150}{5} = 32 \text{ Ohms,}$$

and the current flowing through the dynamo is equal to 5 Ampères, hence its electro-motive force is

$$E_2 = 5 \times 32 = 160 \text{ Volts,}$$

and since the electro-motive force is proportional to the velocity of rotation, the velocity required is

$$= \frac{160}{80} \times 820 = 1640 \text{ turns a minute.}$$

Example 4. When the armature of a certain machine is making 1200 turns a minute, it can send a current of 11 Ampères through an external arc-lamp resistance of 12·5 Ohms. The arc lamps are then replaced by 10 incandescent lamps in simple multiple arc, each of them having a resistance of 125 Ohms. The resistance of the machine is 0·5 Ohm, and it is required to find at what rate the armature must turn so as to maintain a current of 1·2 Ampère through each of these lamps.

Ans. 1309 turns a minute.

Example 5. A current of 10 Ampères is sent through a series of 10 arc lamps, each of 3·8 Ohms resistance, by a dynamo machine whose armature is making 1000 revolutions a minute. The arc lamps are then replaced by 8 incandescent lamps, each of 120 Ohms resistance and arranged in 4 series of 2 each. The armature is then made to turn at 1044 revolutions a minute. The resistance of the machine is 3 Ohms. Assuming that the electro-motive force developed by the machine is proportional to the velocity of rotation, find the strength of current in each lamp.

In the case of the arc lamps the total resistance of the circuit $= 3\cdot8 \times 10 + 3 = 41$ Ohms, and therefore, by Ohm's law the electro-motive force of the machine is

$$E_1 = 41 \times 10 = 410 \text{ Volts.}$$

In the second arrangement the electro-motive force, which is assumed to be proportional to the velocity of rotation of the armature, is

$$E_2 = \frac{1044}{1000} \times 410 = 428 \text{ Volts,}$$

and the resultant resistance of the circuit is

$$= \frac{2 \times 120}{4} + 3 = 63 \text{ Ohms},$$

and ∴ the current flowing through the machine

$$= \frac{428}{63} = 6\cdot 8 \text{ Ampères},$$

and the current through each lamp

$$= \frac{6\cdot 8}{4} = 1\cdot 7 \text{ Ampères}.$$

Example 6. Fifteen incandescent lamps, each of 120 Ohms resistance, are arranged in simple multiple arc between the poles of a dynamo machine whose resistance is 1 Ohm, and which is making 1395 revolutions a minute. When the extra polar circuit consists of 3 arc lamps, each of 3 Ohms resistance, a velocity of 1550 revolutions develops an electro-motive force of 200 Volts. Find the strength of current in each arc lamp, and also in each incandescent lamp.

Ans. Current in arc lamp = 20 Ampères. Current in incandescent lamp = 1·3 Ampère.

Example 7. The poles of a certain dynamo machine whose resistance is 0·08 Ohm, are in one case connected by 4 incandescent lamps in series, and in another case by the same 4 lamps arranged in multiple arc. The resistance of each lamp is 50 Ohms, and the same amount of energy is expended in each case by the driving engine. Compare the amounts of heat developed in the machine.

If H be the total amount of energy developed in the circuit which is the same for both cases, then when the lamps are in series the heat developed in the machine is

$$H_1 = \frac{0.08 \times H}{200 + 0.08} = \frac{H}{2501}.$$

In the second case the resistance of the extra polar circuit is $= \frac{50}{4} = 12.5$ Ohms, and therefore the amount of heat developed in the machine is

$$H_2 = \frac{0.08 \times H}{12.5 + 0.08} = \frac{H}{157.25},$$

$$\therefore \frac{H_2}{H_1} = \frac{2501}{157.25} = \frac{16}{1} \text{ nearly.}$$

N.B.—The student will notice from this result that a dynamo machine should never be let run on *short circuit*, as in that case all the energy developed by the machine is dissipated as heat in the machine itself.

Example 8. In one case 100 incandescent lamps, each of 120 Ohms resistance, were joined up in 25 series of 4 each to the poles of a dynamo machine whose resistance was 2 Ohms. In another case the same lamps were joined up to the same machine in 50 series of 2 each. If the same horse-power was expended in each case by the driving engine find what proportion of the total heat is dissipated in the machine in each case.

Ans. 0·18 and 0·29 respectively.

Example 9. An Edison lamp of 120 Ohms resistance has a current of 0·8 Ampère flowing through it. A wire of 20 Ohms resistance is then placed across its terminals as a *shunt*. If the same current be now divided between the lamp and shunt, by how much will the heat developed in the lamp be diminished?

The current flowing through lamp

$$= \frac{20}{120 + 20} \times 0\cdot 8 = \frac{4}{35} \text{ Ampère,}$$

and if H_1 and H_2 be the quantities of heat developed in the lamp before and after the addition of the shunt, then

$$\frac{H_2}{H_1} = \frac{I_2^2}{I_1^2} = \left(\frac{4}{35}\right)^2 \times \left(\frac{10}{8}\right)^2 = \frac{1}{49}.$$

Example 10. Two Swan lamps, one of which had a resistance of 30 Ohms and the other one a resistance of 20 Ohms, were arranged first in series, and then in multiple arc, and a current of 1·2 Ampère was maintained between the extreme terminals. Compare the amounts of heat developed in each lamp in the two cases.

For the first lamp $\dfrac{H_2}{H_1} = \dfrac{4}{25}$;

For the second lamp $\dfrac{H_2}{H_1} = \dfrac{9}{25}$.

Example 11. A dynamo machine has an internal resistance of ·0046 Ohm, and it supplies a current of ·9 Ampère to each of 1000 incandescent lamps arranged in simple multiple arc. If each lamp has a resistance of 120 Ohms, find the amount of power consumed in the machine and in each lamp, and also the amount of heat developed.

The current flowing through the machine

$= 1000 \times ·9 = 900$ Ampères.

∴ Work consumed in the machine

$= I^2 R = (900)^2 \times ·0046$.
$= 3726$ Watts.

Work expended in each lamp

$= (·9)^2 \times 120 = 97·2$ Watts.

The amount of heat developed in the machine and in each lamp can be readily expressed in terms of the new electro-magnetic unit of heat which has been suggested by Dr. C. W. Siemens. It is the amount of heat which is generated in one second by a current of 1 Ampère flowing through a resistance of 1 Ohm and is called a *Joule*.

In the present case

Heat developed in machine per second = 3726 Joules.
„   „   „ each lamp „   „   = 97·2 Joules.

Example 12. The mechanical equivalent of the amount of heat which is necessary to raise the temperature of one gramme of water 1° C. is $4·2 \times 10^7$ C.G.S. units or ergs. How many grammes of water would be raised 1° C. by the amount of heat represented by one Joule?

One water-gramme degree Centigrade = $4·2 \times 10^7$ ergs.
One Joule = 10^7 ergs.

$\therefore$ One Joule = $\dfrac{\text{one w.g.d. C.}}{4·2}$ = 0·238 w.g.d. C°.

Example 13. The poles of a certain dynamo machine, whose internal resistance is ·125 Ohm, are connected in simple multiple arc by 50 incandescent lamps, each of 100 Ohms resistance, and a current of one Ampère is sent through each lamp. Find the amount of heat expended in each lamp and consumed in the machine per second.   *Ans.* Heat in lamp = 100 Joules.
Heat in machine = 6·25 Joules.

ELECTRIC LIGHT ARITHMETIC. 79

Example 14. Two incandescent lamps, each of 120 Ohms resistance, are connected in multiple arc with the poles of a dynamo machine whose resistance is 4 Ohms, and a current of 1 Ampère is sent through both. The lamps are then placed in series and again a current of 1 Ampère is sent through both. What amount of heat is developed in the machine in each case?   *Ans.* 16 and 4 Joules.

Example 15. The resistance of a certain battery is 13·5 Ohms, and its electro-motive force is 98·5 Volts. Its poles are first short circuited by a thick wire and then are connected by a lamp and leading wires of 11 Ohms resistance. Find the quantity of heat developed in the battery in each case.

Ans. 718·7 Joules and 218·2 Joules.

Example 16. The resistance of each of 1320 Edison lamps, arranged in simple multiple arc, is 140·5 Ohms and that of the armature of the dynamo machine is 0·0042 Ohm. The resistance of the leading wires is 0·01 Ohm and that of the field coils, which are arranged in parallel circuit with the lamps, is 7·067 Ohms. If the machine can convert 142 H.P. into electrical energy, find the amount of heat developed in the field coils. The resistance of the lamp circuit

$$= 0\cdot01 + \frac{140\cdot5}{1320} = 0\cdot1164 \text{ Ohm.}$$

The resistance of the field coils $= 7\cdot067$ Ohms,

$\therefore$ resultant resistance of external circuit

$$= \frac{7\cdot067 \times 0\cdot1164}{7\cdot067 + 0\cdot1164} = 0\cdot1145 \text{ Ohm},$$

and the resultant resistance of whole circuit

$$= 0\cdot0042 + 0\cdot1145 = 0\cdot1187 \text{ Ohm.}$$

$\therefore$ Energy expended in external circuit

$$= \frac{1145}{1187} \times 142 \text{ H.P.}$$

But

$$\frac{\text{energy expended in the field coils}}{\text{,,   ,,   ,, external circuit}} = \frac{\text{resistance of lamp circuit}}{\text{resistance of (lamps + field coils)}}$$

$$= \frac{0\cdot1164}{7\cdot067 + 0\cdot1164} = \frac{0\cdot1164}{7\cdot1834}$$

$\therefore$ Energy expended in the field coils

$$= \frac{0\cdot1164}{7\cdot1834} \times \frac{1145}{1187} \times 142 \text{ H.P.}$$

$$= 2\cdot2195 \text{ H.P.}$$

But the heat equivalent of one H.P. is 746 Joules and $\therefore$ the amount of heat developed per second in the field coils of this dynamo machine is

$$= 2\cdot2195 \times 746 = 1656 \text{ Joules.}$$

ELECTRIC LIGHT ARITHMETIC. 81

Example 17. Find the amount of heat developed in the revolving armature of the above machine.

The energy expended in the armature

$$= 142\left(1 - \frac{1145}{1187}\right) \text{H.P.}$$

$$= 142 \times \frac{42}{1187} = 5\cdot024 \text{ H.P}$$

∴ Heat developed in armature

$$= 5\cdot024 \times 746 = 3748 \text{ Joules.}$$

Example 18. Find the amount of heat which would have been developed in the armature if its resistance had been 0·0041 Ohm instead of 0·0042 Ohm, all the other resistances as well as the total amount of electrical energy in the circuit being the same as before.

Ans. 3662 Joules.

N.B.—The difference in the amount of heat developed in the armature shows that, in the construction of a dynamo machine for the development of strong currents, it is important that the resistance of the armature should be as small as possible.

Example 19. The armature of a certain Edison dynamo machine has a resistance of 0·2 Ohm, and

G

the field coils have a resistance of 41·5 Ohms and are connected in parallel circuit with 70 lamps, each of 140 Ohms resistance, arranged in simple multiple arc. Neglecting the resistance of the leading wires, find the amount of heat developed in the armature and in each of the lamps when the machine is converting 8 H.P. into electrical energy.

Ans. Heat in armature = 566·2 Joules.
Heat in each lamp = 73·6 Joules.

ELECTRIC LIGHT ARITHMETIC.

TABLE OF SQUARES, SQUARE ROOTS, AND RECIPROCALS.

x	x^2	$\sqrt{x}$	$\dfrac{1}{x}$
1	1	1·000	1·0000
2	4	1·414	0·5000
3	9	1·732	0·3333
4	16	2·000	0·2500
5	25	2·236	0·2000
6	36	2·449	0·1667
7	49	2·646	0·1429
8	64	2·828	0·1250
9	81	3·000	0·1111
10	100	3·162	0·1000
11	121	3·317	0·0909
12	144	3·464	0·0833
13	169	3·606	0·0769
14	196	3·742	0·0714
15	225	3·873	0·0667
16	256	4·000	0·0625
17	289	4·123	0·0588
18	324	4·243	0·0556
19	361	4·359	0·0526
20	400	4·472	0·0500

x	x^2	$\sqrt{x}$	$\dfrac{1}{x}$
21	441	4·583	0·0476
22	484	4·690	0·0455
23	529	4·796	0·0435
24	576	4·899	0·0417
25	625	5·000	0·0400
26	676	5·099	0·0385
27	729	5·196	0·0370
28	784	5·292	0·0357
29	841	5·385	0·0345
30	900	5·477	0·0333
31	961	5·568	0·0323
32	1024	5·657	0·0313
33	1089	5·745	0·0303
34	1156	5·831	0·0294
35	1225	5·916	0·0286
36	1296	6·000	0·0278
37	1369	6·083	0·0270
38	1444	6·164	0·0263
39	1521	6·245	0·0256
40	1600	6·325	0·0250

x	x^2	$\sqrt{x}$	$\dfrac{1}{x}$
41	1681	6·403	0·0244
42	1764	6·481	0·0238
43	1849	6·557	0·0233
44	1936	6·633	0·0227
45	2025	6·708	0·0222
46	2116	6·782	0·0217
47	2209	6·856	0·0213
48	2304	6·928	0·0208
49	2401	7·000	0·0204
50	2500	7·071	0·0200
51	2601	7·141	0·0196
52	2704	7·211	0·0192
53	2809	7·280	0·0189
54	2916	7·348	0·0185
55	3025	7·416	0·0182
56	3136	7·483	0·0179
57	3249	7·550	0·0175
58	3364	7·616	0·0172
59	3481	7·681	0·0169
60	3600	7·746	0·0167

x	x^2	$\sqrt{x}$	$\dfrac{1}{x}$
61	3721	7·810	0·0164
62	3844	7·874	0·0161
63	3969	7·937	0·0159
64	4096	8·000	0·0156
65	4225	8·062	0·0154
66	4356	8·124	0·0152
67	4489	8·185	0·0149
68	4624	8·246	0·0147
69	4761	8·307	0·0145
70	4900	8·367	0·0143
71	5041	8·426	0·0141
72	5184	8·485	0·0139
73	5329	8·544	0·0137
74	5476	8·602	0·0135
75	5625	8·660	0·0133
76	5776	8·718	0·0132
77	5929	8·775	0·0130
78	6084	8·832	0·0128
79	6241	8·888	0·0127
80	6400	8·944	0·0125

ELECTRIC LIGHT ARITHMETIC.

x	x^2	$\sqrt{x}$	$\dfrac{1}{x}$
81	6561	9·000	0·0123
82	6724	9·055	0·0122
83	6889	9·110	0·0120
84	7056	9·165	0·0119
85	7225	9·220	0·0118
86	7396	9·274	0·0116
87	7569	9·327	0·0115
88	7744	9·381	0·0114
89	7921	9·434	0·0112
90	8100	9·487	0·0111
91	8281	9·539	0·0110
92	8464	9·592	0·0109
93	8649	9·644	0·0108
94	8836	9·695	0·0106
95	9025	9·747	0·0105
96	9216	9·798	0·0104
97	9409	9·849	0·0103
98	9604	9·899	0·0102
99	9801	9·950	0·0101
100	10000	10·000	0·0100

TABLE OF THE BIRMINGHAM WIRE GAUGE.

B. W. G. No.	Diam. in Inches.	Diam. in Centims.	Circum. in Centims.	Area in Sq. Centims.
1 circ in	1·000	2·540	7·980	5·067
0000	·454	1·153	3·622	1·044
000	·425	1·080	3·393	·916
00	·380	·965	3·032	·731
0	·340	·864	2·714	·586
1	·300	·762	2·394	·456
2	·284	·721	2·265	·408
3	·259	·658	2·067	·340
4	·238	·604	1·898	·287
5	·220	·559	1·756	·245
6	·203	·516	1·621	·209
7	·180	·457	1·436	·164
8	·165	·419	1·316	·138
9	·148	·376	1·181	·111
10	·134	·340	1·068	·091
11	·120	·305	·958	·073
12	·109	·277	·870	·060
13	·095	·241	·757	·046
14	·083	·211	·663	·035
15	·072	·183	·575	·026
16	·065	·165	·518	·021
17	·058	·147	·462	·017
18	·049	·124	·390	·012
19	·042	·107	·314	·009
20	·035	·089	·280	·006
21	·032	·081	·254	·005
22	·028	·071	·223	·004
23	·025	·063	·198	·003
24	·022	·055	·173	·002

www.ingramcontent.com/pod-product-compliance
Lightning Source LLC
Chambersburg PA
CBHW032238080426
42735CB00008B/911